Lab Manual to Accompany Introduction to Veterinary Science

Join us on the web at

agriculture.delmar.com

Lab Manual to Accompany Introduction to Veterinary Science

Second Edition

JAMES B. LAWHEAD, V.M.D.

MEECEE BAKER, Ph.D.

DELMAR
CENGAGE Learning™

Australia • Brazil • Japan • Korea • Mexico • Singapore • Spain • United Kingdom • United States

DELMAR
CENGAGE Learning

Lab Manual to Accompany Introduction to Veterinary Science, 2E
James B. Lawhead and MeeCee Baker

Vice President, Career and Professional Editorial:
Dave Garza

Director of Learning Solutions:
Matthew Kane

Senior Acquisitions Editor:
David Rosenbaum

Managing Editor:
Marah Bellegarde

Product Manager:
Christina Gifford

Editorial Assistant:
Scott Royael

Vice President, Career and Professional Marketing:
Jennifer McAvey

Marketing Director:
Debbie Yarnell

Marketing Coordinator:
Jonathan Sheehan

Production Director:
Carolyn Miller

Production Manager:
Andrew Crouth

Content Project Manager:
Katie Wachtl

Senior Art Director:
David Arsenault

Technology Project Manager:
Mary Colleen Liburdi

Production Technology Analyst:
Thomas Stover

For product information and technology assistance, contact us at
Professional & Career Group Customer Support, 1-800-648-7450

For permission to use material from this text or product,
submit all requests online at **cengage.com/permissions**.
Further permissions questions can be e-mailed to
permissionrequest@cengage.com

Library of Congress Control Number: 2008925691

ISBN-13: 978-1-4283-1227-2

ISBN-10: 1-4283-1227-7

Delmar
5 Maxwell Drive
Clifton Park, NY 12065-2919
USA

Cengage Learning products are represented in Canada by Nelson Education, Ltd.

For your lifelong learning solutions, visit **delmar.cengage.com**

Visit our corporate website at **www.cengage.com**

Printed in Canada
1 2 3 4 5 6 7 12 11 10 09 08

Contents

Preface

This Lab Manual was developed in cooperation with ©Carolina Biological Supply Company in an effort to enhance the technical material found in the accompanying text. The 10 intensive laboratories found in the manual will challenge you as a veterinary science student. Many will take more than one period to complete. Included in both your and the teachers' editions of this manual is a listing of all the laboratories and accompanying materials needed to conduct each exercise. In order to save instructor preparation time, the laboratories selected from ©Carolina Biological Supply Company come complete with all or most all materials. Instructors should be warned to review each laboratory prior to ordering, upon arrival, and then before subsequent usage to ensure the timely arrival, proper storage, and appropriate use. Instructors also should take special note of laboratories with dated supplies.

The sequence of the laboratories mirrors the text, *Introduction to Veterinary Science, 2E*. Since the text was written for those of you in advanced high school and entry-level college courses and who have had previous biology coursework, use of the microscope has been omitted. The first activity begins with examination of the animal cell and is followed by exercises that study tissues. You will then study body systems and topics such as immunology, nutrient analysis, and infectious disease. These investigations will extend the activities found at the end of each chapter in the text and enhance your overall learning experience.

About the Authors

Dr. James Lawhead is a veterinarian in a private mixed animal practice. He works primarily with dairy cattle, dogs, and cats. Dr. Lawhead joined this practice in 1987 following graduation from the University of Pennsylvania, School of Veterinary Medicine. He gained acceptance to veterinary school following completion of his bachelor's degree at Juniata College. Since that time, Dr. Lawhead has become a partner in this practice. He has a special interest in dairy cattle nutrition, providing nutritional service to a number of his clients. Dr. Lawhead enjoys teaching as well and actively supports the local school districts with lectures and demonstrations.

Dr. MeeCee Baker currently works as a senior associate for Wolff Strategies after serving the PA Secretary of Agriculture Dennis Wolff as coordinator of agricultural educator. In addition, Dr. Baker serves as an adjunct professor at the North Carolina State University. She earned both her bachelor's and doctorate degrees from Pennsylvania State University in agricultural education and a master's of science degree from the University of Delaware in agricultural economics. Dr. Baker was the first woman to be elected president of the National Vocational Agriculture Teachers' Association (now known as the National Association of Agricultural Educators). Baker lives on her family farm in Port Royal, Pennsylvania, with her husband, Dr. Robert Mikesell, and her daughter Elizabeth ("Libby") Baker-Mikesell.

Introduction

Laboratories are extremely valuable in illustrating principles taught in textbooks. This Laboratory Manual contains exercises and investigations that complement the *Introduction to Veterinary Science, 2E* text. Performing experiments and observing real life specimens can aid in the learning process. In addition to learning the academic material found in the text, this Lab Manual will assist you, the student, in developing technical skills.

An important part of the laboratory experience is to practice good safety procedures. Sound safety practices are often a matter of using common sense, but several points should be emphasized. It is very helpful to read and understand the experiment prior to beginning. This will help you move through the steps of the procedure safely and efficiently.

Dress appropriately. Loose clothing, dangling jewelry, and long hair have the potential for creating spills and becoming entangled in the equipment. In addition, many of the supplies and chemicals used in the experiments may stain clothing. Protective clothing may be appropriate for many experiments to prevent damage to good clothing.

Wear eye protection. Not every experiment in this manual requires the use of protective eyewear. However, it is important to recognize that many of the chemicals (including many cleaning solutions) used in these laboratory experiments may cause severe damage if splashed into the eyes. Protective goggles are extremely valuable when handling such chemicals. It is important to recognize that the fumes from certain chemicals can be damaging to contact lenses. Ideally, contact lenses should not be worn during the laboratory experiments.

Food and drink are prohibited in the laboratory. This prevents the possibility of contamination of food or accidental ingestion of dangerous chemicals.

Many chemicals are included in these laboratories. Included with the kits are Material Safety Data Sheets that describe the risks, safety, and health considerations involved with each chemical. This material should be reviewed prior to the laboratory and kept readily accessible in the event of problems.

During the laboratories, nothing should be placed in the mouth. This includes pipettes. Pipetting should only be done with a bulb, not by mouth. This technique is more difficult to master, but is important to prevent accidental ingestion of potentially harmful materials.

Wear protective gloves. While handling hazardous chemicals and biologic supplies that potentially could have pathogenic bacteria, latex or vinyl gloves will protect your skin. In addition, you should wash hands thoroughly and frequently. Washing will flush away chemicals and an antibacterial soap will help eliminate pathogens from the skin. It is extremely important to use the utmost safety in any laboratory using bodily fluids. Human blood and other fluids can carry contagious pathogens that could put other individuals at risk. These pathogens could include life-threatening organisms such as HIV and hepatitis. You should be responsible for handling only your own blood. Lancets, test strips, and other items contaminated with blood or bodily fluids should be disposed of safely. If the occasion should arise that one individual must handle another's blood, disposable protective gloves should be worn. Following removal of the gloves, hands should be washed thoroughly. If any human contamination does occur, the area should be washed thoroughly with an antiseptic soap and a medical professional should be contacted. Contact the school administration and nurse to make sure that you are permitted to test your own blood, and ask the school nurse to review techniques involving the handling of blood and the disposal of blood-soaked waste materials.

Keep the work area clean and organized. This helps prevent accidents and keeps the procedure flowing smoothly.

the exercises in this kit only. ©Carolina Biological Supply Company disclaims all responsibility for any other uses of these materials. Included are:

onion bulblets	toothpicks
malachite green	crystal violet
microscope slides	cover slips

Other items needed are:

scalpels or razor blades	forceps
water	droppers
paper towels	microscopes

METHODS

Animal Cells

1. Gently scrape the inside of your cheek or surface of your tongue with a clean toothpick three or four times.
2. Even though nothing may appear on the toothpick, smear it on the center of a clean slide.
3. Discard the toothpick and place a drop of water on the smear.
4. Carefully lower a cover slip over the smear.
5. Examine the slide under the low- and high-power objectives of the microscope. Look for the presence of cheek cells. Note the general size and shape of any you find.
6. Remove the slide from the microscope. Wash and dry the slide and cover slip.
7. Repeat steps one and two using a clean toothpick.
8. Discard the toothpick and allow the slide to dry for one minute.
9. After a minute, place one or two drops of crystal violet on the smear. Allow the slide to sit 20 to 30 seconds, then carefully lower a cover slip over the smear.
10. Place a drop of water near one edge of the cover slip and gently touch a piece of paper towel to the opposite edge. Allow the water to be drawn under the cover slip and clear out the stain.
11. Examine the slide under the low- and high-power objectives of the microscope. Note the effect of the stain on the cheek cells. Look for the cell membrane, cytoplasm, and nucleus. Describe any other objects within or associated with the cells.
12. After examining the slide, wash and dry the slide and cover slip.

Plant Cells

1. Obtain a small onion and cut it lengthwise into quarters.
2. Obtain a clean slide and place one to two drops of water on it.
3. Use forceps to peel the thin skin from the inner side of one of the onion's fleshy scales.
4. Keeping the skin flat, place it on the drop of water and carefully lower a cover slip over it. Gently press on the cover slip with forceps or a similar object to force out any air bubbles.
5. Examine the slide under low and high power.
6. Remove the slide from the microscope. Wash and dry the slide and cover slip.
7. Place one to two drops of malachite green on the center of the slide.
8. Peel a fresh piece of skin from the onion and place it on the drop of malachite green. Allow the skin to sit for one minute.

9. After one minute, carefully lower the cover slip over the skin. Gently press on the cover slip to force out any air bubbles.
10. Examine the slide under the low- and high-power objectives of the microscope. Note the effect of the stain on the skin. Try to identify the cell wall, cell membrane, cytoplasm, and nucleus. Note the cell wall and cell membrane may be hard to differentiate.
11. After examining the slide, wash and dry the slide and cover slip.

DATA

Sketch below what you saw in each of the four prepared slides. Identify any organelles located in each. Ask the instructor for assistance if needed.

Animal cell not stained

Animal cell stained

Plant cell not stained

Plant cell stained

RESULTS

The exercises in this kit are designed to give you a basic introduction to the microscopic study of cells. Animal and plant cells follow similar designs: both have a nucleus and cytoplasm containing various organelles, and both are bounded by a cell membrane.

A major difference between the animal and plant cells is the presence of a cell wall surrounding plant cells. The stained onion skin should plainly show cell walls. Because of the proximity of a plant cell membrane and cell walls, you will probably not be able to distinguish the cell membrane just inside the cell wall.

The stained slide of the cheek cells may reveal several small, darkly stained, rod or spherical shaped bodies. These are bacteria. The bacteria may be associated with the cheek cells or unattached. The bacteria may be in clumps of several cells or scattered singly. The bacteria are naturally present in the mouth and were transferred to the slide with the cheek cells by the toothpick.

Lab 2

Animal Tissue

LAB LINK TO VETERINARY PRACTICE

Mr. Frugal brings his dog Spot into your office. Spot is an 11-year-old Dalmatian. He tells you what a good dog she has been. Over the years, he must have gotten a dozen litters of puppies from her. He was disappointed though; she was just in heat a couple of months ago and she didn't get pregnant. He is concerned because she has quit eating and is losing weight. He asks, "What do you think Doc, is she upset because she didn't get pregnant?"

You examine her and discover that she is running a fever and also has a purulent vaginal discharge. You are concerned that she might have developed a uterine infection. You discuss your concerns with Mr. Frugal and decide to perform a complete blood cell count. The blood sample shows that the white blood cell count is quite elevated, and you prepare a blood smear to examine what kind of white cells are present. This differential count shows that there is an elevated count of neutrophils and monocytes, a distribution that helps to support your diagnosis. Radiographs of the abdomen confirm your diagnosis.

You tell Mr. Frugal that Spot needs to be spayed to save her life. He reluctantly agrees after you assure him that she would never be able to have another litter of pups anyway. You take Spot to surgery. As you make your incision, you incise through several layers of tissue: skin, fat, and connective tissue. You successfully remove the infected uterus and ovaries. The tissue layers are then closed or sutured as you finish the surgery.

Tissues have been a key element in this case. Your knowledge of blood as a connective tissue helped you make this diagnosis. Likewise, your knowledge of tissues allowed for proper handling and surgical technique to create a successful outcome.

INTRODUCTION

The adult animal body consists of many different kinds of tissues. Epithelial, muscle, connective, and nerve tissues are the four major types. The purpose of this set is to introduce these tissue types, and help you identify and understand them.

Staining of the tissue adds color so that you can see the characteristic structures. In some cases, more than one tissue type will be present on a single slide. However, each slide represents a primary tissue, which is the one that will be discussed.

OBJECTIVE

Upon completion of this lab, you should be able to distinguish the four tissue types and discuss their characteristics.

SAFETY GUIDELINES

Comprehensive material safety data sheets are enclosed in the ©Carolina Biological Supply Company laboratory kits. Also, please review the material found at the front of this manual. Observe standard laboratory safety procedures when working with microscopes and microscope slides. Check each slide for cracks or chips prior to usage: broken glass can cut. If you use a natural light source for your microscope, remember direct sunlight focused through a microscope can damage the eyes.

MATERIALS

The materials that accompany the Beginner's Animal Tissues Slide Set from ©Carolina Biological Supply Company #KZ31-1956 are supplied for use with the exercises in this kit only. ©Carolina Biological Supply Company disclaims all responsibility for any other uses of these materials. Included in the kit is an animal tissues slide set. Also needed but not included in the kit is a microscope.

> ⚠ Note
>
> Before looking at these slides, you should be familiar with the general structure and appearance of animal cells. A basic knowledge of body structure and function also would be helpful. For instance, the structures of a bone cross section will be more intelligible if you are familiar with whole bones and skeletons. You also should have some experience in using the microscope. Magnifications of 100X to 400X are suitable for these slides. Always begin with the lowest magnification and move to a higher magnification if necessary. It is not always true that greater magnification produces better viewing. A clear, sharp image at 200X is more useful than a blurred image at 600X. Use indirect, natural light or artificial light.

METHODS

You should observe the following slides to identify the tissue and associated structures detailed below.

Epithelium

Epithelium forms continuous layers of cells that cover or protect internal and external organs. These cells line the digestive and urinary tracts, lungs, glands, and blood vessels. Such exposed surfaces are subject to physical damage, which is constantly being repaired by new cells produced by cell division.

Simple epithelium is only one cell-layer thick. Stratified epithelium is several cells thick. Pseudostratified epithelium appears to be more than one cell thick, but all the cells attach to the same basal membrane (surface for cell attachment). This set contains examples of the first two types.

Stratified Squamous Epithelium

This slide was made by wiping the lining of the mouth, depositing the cells that were rubbed off onto a slide, and staining them. The technique usually separates the cells completely although some clusters of cells may remain together. If you find a folded cell or one standing on edge, note that it is thin and flat; this is the major physical characteristic of squamous (flat) epithelial cells. In stratified squamous epithelium, only the outer layers of cells are flattened. Notice the centrally located oval nucleus and the somewhat granular appearance of the cytoplasm.

Simple Columnar Epithelium

This slide is a cross section of the small intestine showing the villi, the finger-like projections in the cavity of the intestine. Columnar epithelium covers the villi. These cells are taller than they are wide. The nuclei are in the basal ends of the cells, and they tend to be oval and aligned with the long axis of the cell. Notice a red-stained line along the outer surface of these epithelial cells. The staining of microvilli, which are extremely small projections of the epithelial cells, produces this border. They serve, as do the villi, to increase the surface area of the lining of the intestine. This increases the rate of absorption of food nutrients.

The faintly stained, somewhat larger cells are goblet cells, which secrete a mucus that coats the epithelium. This is a glandular function, so the lining of the small intestine is also a glandular epithelium.

Transitional Epithelium

Transitional epithelium is found in structures such as the urinary tract that are subject to stretching. Transitional epithelium is always stratified. The defining feature of this epithelium is that the shape of the cells changes according to whether the tissue is stretched or relaxed. The cells tend to be rounded in a relaxed state, although they may be cuboidal or columnar. When stretched the cells of the surface layer become flattened. However, the flattening is never as great as that of squamous epithelium.

Ciliated Epithelium

Ciliated epithelium lines the inner surfaces of ducts and passageways of many systems in the body including the respiratory and reproductive systems. The face of the cell, which borders the lining of the passageway, is ciliated. The cilia are oriented so they all beat in the same direction, resulting in the flow of material such as mucus in one direction along the surface. In the respiratory system, constant movement of the cilia sweeps dust and other debris from the air passages. In the reproductive system, the cilia help move the sex cells.

Muscle Tissue

Muscle cells are specialized for contraction. Each contractile unit is a muscle fiber. A fiber is a single cell containing one or more nuclei. Contraction of the fiber, an energy-requiring process, results from changes in the proteins forming the myofibrils (long thin rods) within each fiber.

There are three types of muscles: smooth, skeletal, and cardiac. Because of the highly specialized nature of muscle cells, distinctive descriptive terms are used that are specific to these cells. For example, sarcoplasm refers to muscle cytoplasm and sarcolemma refers to the cell membrane and associated material.

Smooth Muscle

This cross section of the digestive tube shows two layers of smooth muscle. In the outer, longitudinal layer, the fibers are cut in cross section. In the inner, circular layer, the fibers are in a long section. Each fiber is an individual spindle-shaped (long and tapered) cell. Myofibrils are not visible. The nucleus is central. Smooth muscle contractions are slow, sustained, and involuntary.

Skeletal (Striated) Muscle

This section through a tongue shows striated muscle fibers in both long and cross section. The fibers are multinucleate. Nuclei lie under the sarcolemma. In cross section, the sarcoplasm appears grainy because you are looking at the cut ends of the myofibrils that fill the cell. The myofibrils align with the long axis of the fiber. Fibers cut in long section show the striations of the myofibrils. Skeletal muscle contractions are fast, not sustained, and voluntary.

Cardiac Muscle

While smooth and striated muscles occur in many parts of the body, cardiac muscle occurs only in the heart. It forms the walls of the heart and its contractions pump blood throughout the body.

In many ways, cardiac muscle is intermediate between smooth and skeletal muscle. Each fiber is a single cell although the cells branch. Individual cells connect at intercalated disks, which show as dark-stained lines. In long section, the fibers appear striated. There is at least one centrally located nucleus per fiber although many cardiac muscle fibers have two nuclei. Cardiac muscle contractions are faster than those of smooth muscle but slower than striated. The contractions are not sustained and are involuntary.

Connective Tissue

Connective tissues are located throughout the body. They support and protect the tissues and organs and hold them in place. They give the body its shape. Connective tissues include both blood cells and bone. All these diverse tissues develop from a specific cell type that arises early in the development of the embryo.

Adipose Tissue

Adipose tissue, commonly referred to as fat, occurs in a layer under the skin. It insulates against heat loss and gives a smooth outline to the body. Adipose tissue also lines the eye sockets, forming a protective shock-absorbing layer. In times of famine, it is vital as storage tissue.

Looking at adipose tissue under a microscope is like looking down a layer of soapsuds. Fat is stored as a droplet that fills most of the cell. The process of making this slide has dissolved the fat, leaving an empty space. Under high-power magnification, you can see the nucleus and the cytoplasm pushed into a thin layer against the cell membrane. The cells increase in size as they store more fat.

Blood

Blood has two components: liquid and cellular. Blood cells, unlike the cells of other tissues, do not connect to each other or to any substrate. Thus, blood is a fluid.

As a beginning activity, it will be enough for you to identify red and white cells. As an extra challenge, you may try to identify various kinds of white cells. For this purpose, use color illustrations if possible.

There are three categories of formed elements: red blood cells (erythrocytes), white blood cells (leukocytes), and platelets.

Red Blood Cells. These are disk-shaped cells without nuclei. Red pigment called hemoglobin colors the cytoplasm. Red blood cells are by far the most numerous of the blood cells.

White Blood Cells. These nucleated cells have clear (transparent) cytoplasm, which may or may not contain dark-staining granules.

Granulocytes and agranulocytes are the two major types of white blood cells. Granulocytes have prominent cytoplasmic granules and their nuclei are highly lobed. Agranulocytes have few or no cytoplasmic granules and round or kidney-shaped nuclei.

There are three groups of granulocytes: neutrophils, eosinophils, and basophils. Neutrophils are by far the most common of the white blood cells. Over half the white blood cells on the slide are neutrophils. The nucleus consists of two to five lobes connected by thin strands. Granules stain pale pink. The nucleus of an eosinophil is usually two-lobed. Cytoplasmic granules stain red. You may find two or three eosinophils on the slide. Basophils are similar to eosinophils except the granules stain blue. Basophils are the least common of the white blood cells, and you may not find one on this slide.

Monocytes and lymphocytes are two types of agranulocytes. Monocytes are the largest cells found in blood. They have a diameter two to three times that of red blood cells. Their nuclei may be round, oval, kidney-shaped, or lobed. There should be three or more on the slide.

Lymphocytes are second only to neutrophils in abundance. The nucleus is large and usually round although it may be kidney-shaped. The cytoplasm usually forms a thin layer around the nucleus.

Platelets. These are the smallest of the formed elements of the blood. Strictly speaking, platelets are not cells but fragments from large cells in the bone marrow.

Bone

Bone is a hard, rigid tissue. Its hardness is due to the presence of calcium phosphate and calcium carbonate salts deposited by the bone cells (osteocytes).

This preparation was made by attaching a thin cross section of bone to a slide and then grinding the bone even thinner, leaving only the extracellular materials. The bone tissue is organized around haversian canals, tiny tubes through the bone that in life, contain blood vessels. Arranged in more or less concentric circles around each canal are the lacunae with what seem to be radiating cracks. In life, a bone cell fills each lacuna and strands of cytoplasm extend into the "cracks." In this way, the bone cells connect to each other and (directly or indirectly) to the blood vessel in the canal. These structures make up the haversian system, which, in long section, would look like a cylinder of bone.

Fibrocartilage

The gristle found in meat is actually cartilage, a rigid connective tissue that provides support and protection. There are three types of cartilage: hyaline, elastic, and fibrocartilage. Fibrocartilage is extremely tough. It acts as a shock absorber in high stress areas such as the disks between joints of the spine.

Under the microscope, fibrocartilage shows a matrix of fibers that is characteristic of all cartilage. Scattered about in the fibers are individual cartilage cells (chondrocytes). Blood vessels are absent, so oxygen and nutrients must diffuse into cartilaginous tissue. As a result, cartilage is slow to recover from injury.

Nervous Tissue

The nerve cell or neuron is specialized for the conduction of an electrochemical event called a nerve impulse. The impulse initiates when the neuron responds to a stimulus. Long processes called nerve fibers enable a neuron to conduct impulses over distances of up to several meters. Dendrites are processes that conduct impulses toward the cell body and axons conduct impulses away from the cell body. Further, sensory neurons conduct impulses toward the central nervous system while motor neurons conduct impulses away. Normally, a neuron conducts impulses in only one direction.

A nerve, whether sensory or motor, consists of nerve fibers and connective tissue. Nerve fibers make up most of the white matter of the brain and spinal cord, while the gray matter contains nerve fibers as well as large numbers of nerve-cell bodies. A mature neuron loses the ability to divide. This limits the ability of nervous tissue to repair itself, and injuries are often permanent.

Spinal Cord

Examine this cross section under the lowest power of your microscope or under a stereomicroscope. Locate the areas of white and gray matter (due to staining, these areas are no longer white and gray). Change to a higher power and examine the white matter. Here you will see cross sections of nerve fibers. The dark central core is the axon. The clear covering is the myelin sheath. It is this sheath that makes the white matter white. Nerve fibers in the gray area lack sheaths.

In the gray area, you will find large nerve cells. Each has a central nucleus that stains almost the same color as the cytoplasm and is, therefore, hard to see. In contrast, it is easy to see the nucleolus because it stains heavily.

Neuron

Smearing tissue from a spinal cord onto the slide and staining it made this slide. You will see numerous small dark-stained cells. These are neuroglia. They are found only within the central nervous system, and they support and otherwise aid the nerve cells. You will also see several motor nerve cells. Some of these should be free enough of overlying material for you to see the nerve processes extending from the cell body.

Nerve

This is a cross section of a nerve. Notice that the nerve fibers are in bundles surrounded by connective tissue. Connective tissue also extends between and around the individual nerve fibers. Under high power, you will see cross sections of nerve fibers with myelin sheaths similar to those you saw in the white matter of the spinal cord.

Nerve Fibers

On this slide, a length of nerve was teased apart to show the individual nerve fibers. The myelin sheath is stained black. Schwann cells produce the sheath. The indentation between the sheaths of successive Schwann cells is the node of Ranvier. By focusing with the fine adjustment on high power, you can see the nerve fiber within its sheath.

DATA AND RESULTS

Since this laboratory exercise was experiential and not experimental in nature, you do not need to record data or results.

Lab 3

Contraction of Glycerinated Muscle with ATP

INTRODUCTION

Muscle tissue is made of fibers formed by the fusion of cells during development. A single muscle fiber, barely visible to the unaided eye, has many nuclei that lie close to its outer membrane. Each fiber contains hundreds of long, thread-like structures called myofibrils, arranged in parallel. About 75% of a muscle's total volume is made up of myofibrils. Myofibrils are the structures that carry out muscle contraction.

Under a microscope, myofibers look striated (striped) with a repeating pattern of bands and lines perpendicular to the length of the fiber. The banded pattern is caused by an organized, parallel arrangement of protein filaments within the myofibrils. There are two types of filaments in a myofibril: thick filaments composed of the protein myosin and thin filaments composed of the protein actin. Actually, both filaments are quite thin on a human scale, but myosin filaments are thicker than actin filaments.

The actin and myosin filaments arrange in a very organized manner. The filaments overlap in an orderly, repeated manner, creating units called sarcomeres. When many filaments are bundled in a cylinder, the repeated overlapping pattern of filaments results in the banded pattern seen under the microscope. Letters of the alphabet designate the different bands in the visible pattern and they correspond to different segments of the sarcomere. When observing glycerinated muscle fibers with a compound microscope, these bands should be visible.

Muscle contraction occurs through the interaction of the actin and myosin filaments in the sarcomeres. When a muscle contracts, the myosin crossbridges bind to the actin filaments in a manner that causes the acting filaments to be pulled together across the H zone. Under the light microscope, the A and I bands become narrower, and the overall width of the sarcomeres decreases.

At the molecular level, thin filaments are composed of two chains of identical actin monomers twisted around each other in a double helix like two twisted strands of pearls. Thick filaments are composed of hundreds of myosin molecules, each one a long rod with a globular head. In the fiber, the myosin molecules are arranged so that the rods lie alongside one another and the globular heads protrude away from the fiber.

For a muscle fiber to contract, the myosin heads must first be activated by ATP. One molecule of ATP binds to a myosin head and is hydrolyzed to ADP and inorganic phosphate (P_i). Both ADP and P_i remain bound to the myosin head, and the energy released from ATP hydrolysis is transferred to the myosin head as well. The myosin head is now activated. Imagine holding one end of a thin plastic ruler and pulling back on the other end so that the ruler bends. When you release the ruler, it will spring back to its original shape. The activated myosin head is rather like the bent ruler. When the myosin head binds to the actin filament, its energy is released and the myosin head springs back, carrying the bound actin filament with it. This movement causes the muscle fiber to contract. Each one of the actin monomers has a binding site for myosin.

After the myosin head has sprung, it can interact with a new molecule of ATP. When ATP binds to the myosin head, it releases the actin fiber, ADP, and P_i. The myosin head is now reactivated and the cycle can begin again. If no ATP is available to reactivate the myosin, the actin/myosin complex remains locked together, and the muscle cannot relax. When an animal dies, its cellular ATP stores are depleted and all its muscles lock. This locked condition is called rigor mortis. In living animals, muscles resume their normal shapes (relax) after contraction because opposing muscles pull them.

The just-described picture of contraction in living muscle is incomplete because it omits the role of nerve signals in instigating contractions. Muscle fibers do not normally contract without appropriate nerve signals because a control mechanism prevents them from doing so. The control mechanism works through two regulatory proteins called tropomyosin and troponin, which form a complex lying along the actin filament. The troponin/tropomyosin complex blocks the myosin binding site on the actin fiber, preventing myosin from binding to actin and causing contraction.

The signal for muscle contraction is a nerve impulse to the muscle fiber that results in intracellular release of calcium ions. The calcium ions induce contraction by binding to troponin. When Ca^{++} binds to troponin, the shape of the troponin/tropomyosin complex is altered. This change in configuration allows activated myosin crossbridges to bind to the actin and release their energy as motion. Therefore, in a normal muscle fiber rich in ATP, the myosin heads are activated and ready to cause a contraction, but they cannot do so until a nerve impulse releases Ca^{++}. The contraction cycles will continue as long as intracellular Ca^{++} concentration is high and as long as ATP is available. When Ca^{++} levels fall, Ca^{++} is released from the troponin molecules and the troponin/tropomyosin complex again blocks the binding sites on the actin fiber.

The glycerinated muscle system is different from muscle in living tissue. The glycerination process removes ions and ATP from the tissue, and disrupts the troponin/tropomyosin complex so that the binding sites on the actin fibers are no longer blocked. No Ca^{++} is needed to induce contraction. However, no ATP is present in the glycerinated tissue, so the myosin heads are not activated. You will be experimenting with adding ATP and ions to the glycerinated tissue to initiate contraction. When contraction occurs, you will be able to see the change in length of the dissected muscle tissue. After the muscle is contracted, it will not relax because there is no opposing muscle to stretch it out.

This kit from ©Carolina Biological Supply Company #KZ20-3525 provides a strip of glycerinated skeletal muscle tissue from which muscle fibers (myofibers) can be dissected for study. Using a compound microscope, observe the striated pattern in the fibers and measure the length of the relaxed sarcomeres. Then induce muscle contraction by adding ATP, observe the contraction, and measure the post-contraction width of the sarcomeres. You should also compare the effects of adding ATP plus potassium and magnesium ions, ATP alone, and the ions alone. Shortening of the fibers can be seen macroscopically and measured without use of a compound microscope.

OBJECTIVE

Upon completion of this lab, you should be able to detail muscle structure and the role of ATP in muscle contraction.

SAFETY GUIDELINES

Comprehensive material safety data sheets are enclosed in the ©Carolina Biological Supply Company laboratory kits. Also, please review the safety guidelines found at the front of this manual.

MATERIALS

The materials that accompany the laboratory kits from ©Carolina Biological Supply Company #KZ20-3525 are supplied for use with the exercises in this kit only. ©Carolina Biological Supply Company disclaims all responsibility for any other uses of these materials. Included supplies are:

glycerinated skeletal muscle strip tied to a stick in a tube of 50% glycerol	dropper vial 0.25% ATP plus 0.05 M KCL plus 0.001 M $MgCl_2$ in distilled water
dropper vial 0.25% ATP in distilled water	dropper vial 0.05 M KCL plus 0.001 M $MgCl_2$ in distilled water

Also needed are:

sharp scissors	teasing needle and watchmaker forceps, or glass needles
petri dishes	
dropping pipettes	millimeter scale
microscope slides and coverslips	compound microscope (optional)
dissecting microscope (optional)	

⚠ Storage Note

Glycerinated muscle preparations can be stored in a freezer indefinitely. Store the solutions of ATP and salts in the refrigerator (4°C to 6°C). Remove the materials from storage just before use. Solutions containing ATP should be used within 10 days of receiving the kit.

METHODS

1. You and/or your work group will need:
   teasing needle and watchmaker forceps, or one pair of glass needles
   dropping pipette
   petri dish with glycerol and skeletal muscle tissue
   five microscope slides and two coverslips
   millimeter scale
   compound or dissecting microscope (optional)

> ⚠ Note
> All glassware and dissecting tools should be cleaned thoroughly and rinsed well in distilled water before use.

2. Remove the stick to which the strip of skeletal muscle has been tied from the test tube. This strip consists of several hundred muscle fibers.
3. Pour the glycerol into a petri dish.
4. Cut the muscle into pieces about 2 cm in length and drop these in the glycerol in the petri dish. One piece of muscle tissue is sufficient for you or your team, although individual work is preferable.
5. Distribute some of the glycerol and one piece of the muscle tissue into a petri dish.
6. Unused muscle tissue should be returned to the freezer in 50% glycerol.
7. Place the petri dish containing a segment of skeletal muscle tissue on the stage of a dissecting microscope (if available).
8. Looking through the microscope, use glass needles or a teasing needle and gently tease the segment into very thin strands. You will see the clearest results with single muscle fibers, but these are difficult to obtain: the thinnest strand you are likely to get is a group of two to four fibers. Strands of muscle exceeding 0.2 mm in cross-sectional diameter are too thick to be used.
9. Mount a thin strand on a microscope slide with a coverslip.
10. Examine the strand under low and high magnification of a compound microscope.
11. Note the striations in the myofibers.
12. Transfer three or more of the thinnest strands to a tiny amount of glycerol on a second microscope slide.
13. Lay the strands out straight and parallel to each other. Do not cover them.

> ⚠ Note
> The amount of glycerol needed depends on the heat of the microscope lamp and the length of exposure to the heat. With no appreciable heat, the glycerol that adheres to the strand of fibers is sufficient. The less glycerol used, the easier the fibers are to measure.

14. Place the slide under a dissecting microscope and measure the length of the strands with a millimeter scale held beneath the slide.
15. Record these lengths.
16. Flood the strands with several drops of the solution containing ATP plus potassium and magnesium ions.
17. Observe the reaction of the fibers.

⚠️ Note

It is essential to avoid cross contamination between the ATP and salt solutions. Such contamination will lead to ambiguous experimental results.

18. After 30 seconds or more, remeasure the strands and calculate the degree of contraction. Has the fiber changed in width?
19. Remove one of the contracted strands to another slide.
20. Examine it under a compound microscope and compare the fibers with those seen in Step 11.
21. Note the differences.
22. Repeat the experiment using clean slides, new myofibers, and the solutions of ATP alone and salts alone.

	Starting Measurements		Final Measurements	
	Length	Width	Length	Width
ATP + KCL + MgCl$_2$				
ATP alone				
KCL + MgCl$_2$				

23. What conclusions may be drawn from your results?

DATA

Write your findings in the table below.

RESULTS

The speed and extent of the muscle contraction you will observe are influenced by the amount of glycerol on the slide, the concentration of active ATP, the ions present, and the width of the dissected muscle strand. Under favorable conditions, myofibers can be expected to contract to almost 50 percent of their starting length within 10 seconds. Although it is easier using a dissecting microscope, you can tease the muscle strip into satisfactory strands and measure their concentration using the unaided eye.

1. Calculate the percent change in length of each muscle fiber. Record any change in width as well.

this kit only. ©Carolina Biological Supply Company disclaims all responsibility for any other uses of these materials:

simulated urine low	simulated urine normal
simulated urine high	simulated urine unknown A
simulated urine unknown B	urine containers
urine hydrometer and jar	glass vials, dropping pipettes
jumbo pH strips, wide range	Clinitest tablets with chart
biuret reagent	marking pen
student guides	

Items needed but not provided in the kit are:

10 ml graduated cylinder

METHODS

Part One

Testing Normal and Abnormal Urine

In the first part of this lab, you will test three simulated urine specimens for color, pH, specific gravity, glucose, and protein. Urine low is abnormally low (negative), urine normal is normal, and urine high is abnormally high (positive) on all tests. In the second part, you will test unknown abnormal simulated urines and determine the probable disorders they represent.

Color. Record the color of each simulated urine (Low, Normal, and High) in Table 4–1. (Table 4–2 lists urine color and possible causes).

Table 4–1. Data

Urine Test	Low	Normal	High	Urine A	Urine B
Color					
pH					
Specific Gravity					
Base					
Calibration Factor					
Adjusted					
Glucose					
Protein					

pH.

1. Obtain three pH indicator strips and label each to match the containers (L, N, and H).
2. Test each simulated urine (Low, Normal, and High) by dipping the appropriate pH strip into it three consecutive times.
3. Shake off the excess liquid and compare the color of the strip to the colors on the pH chart. The color that closely matches your strip corresponds to the pH of the simulated urine.
4. Record each pH in Table 4–1. The pH of normal urine averages 6.0, which is slightly acidic. (Foods and diseases that can affect urine pH are listed in Table 4–3.)

Specific Gravity.

1. The instructor will either calibrate the hydrometer or provide you with a calibration factor. Record this factor in Table 4–1.
2. Obtain the urine hydrometer and jar, and rinse both pieces well. Fill the jar three-quarters full of urine low. Place the hydrometer in the jar so that it is not touching the sides and read the level of urine on the hydrometer scale. Record the measured (base) urine specific gravity in Table 4–1.
3. Your instructor will instruct you to either add to or subtract from your reading of the calibration factor. Record the adjusted urine specific gravity in Table 4–1.
4. Remove the hydrometer and pour the simulated urine sample back into the container.
5. Rinse the hydrometer and jar well.
6. Repeat Steps 2 to 5 with urine normal and urine high. Record your results in Table 4–1.

Glucose.

1. Using a dropping pipette, place five drops of urine into the glass vial.
2. Rinse the dropper thoroughly with water and add 10 drops of water to the vial.
3. Drop the Clinitest tablet into the vial. Place the vial on the tabletop and observe the reaction.
4. After the reaction has stopped, wait 15 seconds. Shake the vial gently to mix the contents. *Caution: Do not allow the contents of the vial to contact your eyes or skin.* Compare the color to the Clinitest Color Chart.
5. Record the results of the test for glucose (positive or negative in Table 4–1). Glucose (sugar) should not be detected in normal urine. The presence of glucose usually indicated diabetes mellitus, a severe metabolic disorder due to defective carbohydrate utilization. (See Table 4–3 for causes of glucose in the urine.)

Protein.

1. Using a 10-ml graduated cylinder, add 1 ml urine to a clean glass vial.
2. Rinse the graduated cylinder and measure 2 ml biuret reagent. Note the pale blue color of the biuret reagent.
3. Add the biuret reagent to the urine vial.
4. Gently swirl the vial to mix the contents. After 10 minutes, hold the test tube against a white background and observe the color. A color change from the light blue to pale violet indicates the presence of protein.
5. Record the results of the test for protein (positive or negative) in Table 4–1. Normally, protein should not be detected in urine. (Factors associated with the excretion of protein into the urine are listed in Table 4–3.)

RESULTS

Refer to Tables 4–1, 4–2, and 4–3, and compare Table 4–1 results to the data in Tables 4–2 and 4–3.

Table 4–2.   Urine Colors and Possible Causes

Color	Diet	Drugs	Disease
Light yellow/amber	Normal	Phosphate, carbonate	Uncontrolled diabetes mellitus
Clear/light yellow	Alcohol	Antibiotics	Bilirubin from obstructive jaundice
Red to red-brown	Beets	Laxatives	Hemoglobin in urine
Smoky red	Beets	Anticonvulsives	Unhemolyzed red blood cells from the urinary tract
Dark wine	Beets	Anti-inflammatory drugs	Hemolytic jaundice
Brown-black	Rhubarb	Anti-depressants	Melanin pigment from melanoma
Green	Green food dyes	Diuretics	Bacterial infection

Table 4–3.   Abnormal Urinalysis Results and Possible Causes

Test Result	Diet	Disease
Low pH (<4.5)	High protein diet, cranberry juice	Uncontrolled diabetes mellitus
High pH (>8.0)	Diet rich in vegetables, dairy	Severe anemia
Low specific gravity (<1.010)	Increased fluid intake	Severe renal damage
High specific gravity (>1.025)	Decreased fluid intake, loss of fluids	Uncontrolled diabetes mellitus, severe anemia
Glucose present	Large meal	Uncontrolled diabetes mellitus
Protein present	A high protein diet (In veterinary medicine, a high protein diet is not expected to result in protein in the urine)	Severe anemia

Then answer the following questions:

1. Are your urinalysis results normal?

2. The following abnormal results were obtained from a patient's urinalysis:

   Color, very light yellow; pH, 3.0; Specific Gravity, 1.040

   Glucose, positive; Protein, negative

   Name a disease that could cause these results.

OBJECTIVE

Upon completion of this lab, you should be able to explain the role of enzymes in digestion, and the effects of temperature, pH, and concentration on effectiveness.

SAFETY GUIDELINES

Comprehensive material safety data sheets are enclosed in the ©Carolina Biological Supply Company laboratory kits. Also, please review the safety guidelines found at the front of this manual. Take extra caution with the warm water.

MATERIALS

The materials that accompany the Digestion BioKit® from ©Carolina Biological Supply Company #KZ20-2340 are supplied for use with the exercises in this kit only. ©Carolina Biological Supply Company disclaims all responsibility for any other uses of these materials.

dialysis tubing	pepsin solution 0.4%
test tubes	artificial gastric juice
pipettes	hydrochloric acid 0.2%
plastic scoops/stirrers	pancreatic solution
glass-marking pencil	pancreatin (powder)
albumin (powder)	biuret solution
starch solution 5% (powder)	Benedict's solution
sugar solutions 15% glucose (powder)	iodine-potassium-iodide
litmus milk solution (powder)	distilled water

Other items needed are:

incubators or warm water baths	laboratory thermometers
250 ml beakers	test tube rack if incubator is used

METHODS

Gastric Digestion

Glands in the stomach wall secrete gastric juices containing hydrochloric acid and the enzyme pepsin.

1. With a glass marking pencil, label three test tubes and number them 1, 2, and 3.
2. Place one scoop of albumin (rich in protein) in each.
3. Add the following: to test tube 1, 5 ml of 0.4% pepsin solution; to test tube 2, 5 ml of the 0.2% hydrochloric acid; and to test tube 3, 5 ml of artificial gastric juices (pepsin and hydrochloric acid).
4. Stir each test tube to mix the contents well.
5. Place the test tubes in an incubator or warm water bath at 37°C to 40°C (99°F to 104°F) for one hour, then remove the test tubes.
6. Add two to three drops of biuret reagent to each test tube and shake.
7. Record color changes below under Data.

DATA

Test Tube	Color
1	
2	
3	

RESULTS

Biuret reagent reacts with the products of gastric protein digestion (peptides and proteases) to give a pink-violet color.

Whole proteins give a purple color. In which test tubes did the protein digestion occur?

Intestinal Digestion—A

Digestion begins in the mouth and stomach and is completed in the small intestine. Secretions of the pancreas contain enzymes, which complete the digestion of carbohydrates, proteins, and lipids (fats and oils).

1. With a glass marking pencil, label three test tubes with your initials and number them test tube 1, 2, and 3.
2. Add 5 ml of 5% starch solution to each.
3. To test tube 1, add 5 ml of pancreatic solution; to test tube 2, add 5 ml of water; and to test tube 3, add 5 ml of boiled pancreatic solution.
4. Shake gently to mix contents.
5. Place the test tubes in an incubator or warm water bath at 37°C to 40°C (99°F to 104°F) for 30 minutes, then remove the test tubes.
6. Add two to three drops of iodine-potassium-iodide solution to each test tube.
7. Shake gently to mix the contents.
8. Record color changes below under Data.

DATA

Test Tube	Color
1	
2	
3	

RESULTS

The presence of starch is indicated by a light brown color.

What happened to the starch in test tube 1?

What is the product of starch breakdown?

Why do the results in test tube 3 differ from the results in test tube 1?

Intestinal Digestion—B

1. With a glass-marking pencil, label two test tubes with your initials and number them 1 and 2.
2. To each, add 10 ml of litmus milk solution.
3. To test tube 1, add one scoop of pancreatin powder; to test tube 2, add a few drops of distilled water.
4. Shake gently to mix the contents.
5. Place the test tubes in an incubator or warm water bath at 37°C to 40°C (99°F to 104°F) for one hour.
6. Remove the test tubes and observe the color changes.
7. Record color changes below under Data.

DATA

Test Tube	Color
1	
2	

RESULTS

Acid production is indicated by a color change to dark pink-purple.
Milk contains proteins and lipids.

Why would the digestion of these nutrients cause an increase in acidity?
(*Hint:* What are the products of protein and lipid digestion?)

Answer the following questions.
    Was there a color change?

    If so, what was the color?

Absorption in the Small Intestine

1. Soak a piece of dialysis tubing in water for 30 seconds, then rub it between your thumb and index finger to open it.
2. Knot one end and half fill with an equal mixture of starch (5%) and sugar (15%).
3. Tie off the other end.
4. Rinse the outside of the tubing in tap water and wipe it dry to remove traces of nutrients.
5. Place it in a 250-ml beaker of water for 30 minutes.
6. Remove the tubing.
7. In two test tubes, pour 5 ml of the water from the beaker.
8. Add 20 drops of the Benedict's solution to one end of the tube and place it in a boiling water bath for five minutes.
9. Record color changes below under Data.

DATA

Test Tube	Color
1	
2	

RESULTS

In the presence of a simple sugar, the Benedict's solution will turn red.

The absence of color indicates the absence of sugar. Is sugar present in the solution?

10. Add two to three drops of iodine-potassium-iodide to the second test tube.
11. Record color changes below under Data.

DATA

Was there a color change?

If so, what was the color?

RESULTS

The presence of starch is indicated by a blue-black color.

Is starch present in the sample?

Lab 6

Starch Breakdown

LAB LINK TO VETERINARY PRACTICE

Abe Sint Minded, the hired man for a local farm, calls you late one afternoon and sounds like he is panicking. "Doc, you've got to come quick. I ground up the corn this morning just like the boss said and I ran it into the feed bin. But I forgot to close the hatch and it all ran down into the heifer pen. Now there are two heifers down and moaning, and about five or six others that are bloating. Hurry . . . you need to fix them before the boss gets back!"

When you arrive at the farm, Abe begins to ask questions as you examine and treat the heifers. "Doc, these heifers eat corn all the time. Why did eating more make them sick?"

You explain to Abe that corn is high in starch. The microorganisms in the rumen break down the starch into sugars. The metabolism of all this sugar then results in the production of acids that make the heifer sick.

Abe stares at you blindly and says, "I don't really understand that, but you are going to make them better, aren't you?" You explain to Abe that you are going to try to minimize the effects of all that acid, but that some of these heifers are extremely sick and may not make it. The heifers that ate much less are likely to recover without problems.

This lab will prove to Abe how starch is broken down and that sugars are truly produced.

INTRODUCTION

Alpha amylase is an enzyme found in our saliva, which breaks down starch molecules into smaller molecules called dextrins. The pancreas also produces amylase which helps to complete the digestion. Complete digestion of starch into sugars allows the digestive tract to absorb the nutrients for use in the body.

OBJECTIVE

Upon completion of this lab, you should be able to use enzymes to convert corn starch to sugar dextrose.

SAFETY GUIDELINES

Comprehensive material safety data sheets are enclosed in the ©Carolina Biological Supply Company laboratory kits. Also, please review the safety guidelines found at the front of this manual.

Use extreme caution when using the heating devices.

MATERIALS

The materials that accompany the Starch Breakdown BioKit® from ©Carolina Biological Supply Company #KZ20-2335 are supplied for use with the exercises in this kit only. ©Carolina Biological Supply Company disclaims all responsibility for any other uses of these materials. Included in the kit are:

bacterial amylase   fungal alpha amylase
amyloglucoside   corn starch
iodine-potassium-iodide reagent   pH paper
dropper bottle HCl, 1N   stirring rods
medicine droppers   filter paper

Not included in the kit but needed are:

400 ml beakers
200 ml Erlenmeyer flasks
100 ml graduated cylinders
hot plate or gas burner
20 × 150 mm test tubes
thermometers
5 ml pipettes

METHODS

Gelatinization

Heating the starch in water brings about gelatinization of starch molecules. The starch is insoluble at room temperature, but with increased heat, the hydrogen bonds of the starch molecules weaken. The weakened bonds allow the starch molecules to absorb water and to swell, giving a gel-like consistency to the solution.

1. To a 200-ml Erlenmeyer flask, add 100 ml of water and 10 g of corn starch.
2. Gently heat this mixture, stirring continuously, until it begins to thicken.
3. When the mixture thickens, remove from the heat and cool under running water until lukewarm.
4. Remove two drops of the thickened starch and combine with one drop of iodine-potassium-iodide reagent to test for starch.
5. Record color under Data.

DATA

Color:

RESULTS

A dark blue color indicates the presence of starch.

Liquefaction

When gelatinized starch is hydrolyzed by bacterial amylase, the glucose polymer chains of the starch molecules are broken at random, resulting in various sized units call dextrins.

    The breaking of the starch molecules into dextrin destroys the viscosity of the solution. This is called liquefaction.

1. To the flask of starch mixture, add 3 ml of bacterial amylase and shake well to ensure mixing.

2. Allow the solution to sit for 10 minutes, noting any change of viscosity of the mixture.
3. After 10 minutes, filter through No. 201 filter paper, collecting about 60 ml of filtrate. Test two drops of this filtrate with the iodine-potassium-iodide reagent.
4. Record color under Data.

DATA

Color:

RESULTS

A dark blue color indicates the presence of starch.

Saccharification

If the dextrins are now treated with fungal alpha amylase, specific linkages are broken into units composed of two glucose molecules each. This is sugar maltose. The enzyme amyloglucosidase also acts at specific linkages to split molecules into single glucose units. This is sugar dextrose. The action of both alpha amylase and amyloglucosidase is called saccharification.

1. Adjust the pH of the filtrate to 4.7 using the test paper and one or two drops of 1N HCl.
2. Mark one test tube A and one test tube B, and add one-half of the solution to each tube.
3. To test tube A, add 1 gram of the alpha amylase and mix.
4. To test tube B, add 3 ml of the amyloglucosidase and mix.
5. Make a water bath by adding 100 ml of water to a 400-ml beaker and heat to 70°C (158°F).
6. Place test tubes A and B in the water bath and incubate for 30 minutes at 70°C.
7. Remove the test tubes and allow them to cool to room temperature.
8. Test the solutions in test tubes A and B with the iodine-potassium-iodine reagent.
9. Test each solution by taste as well.
10. Record the results under Data.

DATA

Test Tube	Color	Taste
A		
B		

RESULTS

The sugar dextrose should have a much sweeter taste than the sugar maltose.

Lab 7

Immunology Test

INTRODUCTION

In 1948, S. D. Elek and O. Ouchterlony independently reported on a technique of antigen-antibody analysis based on diffusion and precipitation in agar. This technique, known as Ouchterlony double diffusion, is utilized in the Immunology BioKit® from ©Carolina Biological Supply Company.

Two or more closely spaced wells are cut in agar. One well is filled with a serum containing an antigen and the other well(s) is filled with sera containing antibody. As the molecules of antigen and antibody are soluble in the liquid phase of the agar, they diffuse through the agar until they meet. When molecules of antigen and antibody have diffused together, one of two things happens. If the antibody is not specific for that particular antigen, they diffuse past each other, as happens with the molecules of red and green dye in the "Diffusion and Precipitation in Agar" exercise. However, if the antibody is specific for that particular antigen, they react to form an antigen-antibody complex that is insoluble in the liquid phase of the agar and thus precipitates. The precipitate becomes visible as a line or layer in the agar between the well containing the antigen and the well containing the antibody.

Often, the precipitation line will be curved rather than straight. This indicates that one group of molecules has diffused faster than the other. For example, if the precipitin line bends away from the antigen and toward the antibody, this means that the molecules of antigen have diffused faster than the molecules of antibody.

Although several factors control the relative rates of diffusion of antigen and antibody through agar (solubility of the molecules, absorption of molecules on the gel phase of the agar, and the like), the relative rates of diffusion are generally considered to be related to the molecular weights of the molecules. Thus, large heavy molecules diffuse more slowly than small light molecules. Applying this interpretation to the example given, it is likely that the antigen molecules are smaller and lighter than the molecules of antibody.

In this exercise, you will study a technique of immunology and apply it in a test for food purity. Immunology, the study of an organism's response to a foreign organic substance (antigen), has many medical, biochemical, and bacteriological interrelationships.

OBJECTIVE

Upon the completion of this lab, you should be able to follow basic immunology test techniques.

SAFETY GUIDELINES

Comprehensive material safety data sheets are enclosed in the ©Carolina Biological Supply Company laboratory kits. Also, please review the safety guidelines found at the front of this manual. Care should be taken when handling raw meat products. Participants should take care to wash hands thoroughly after handling.

MATERIALS

The materials that accompany the Immunodetective BioKit® from ©Carolina Biological Supply Company #KZ-20-2100 are supplied for use with the exercises in this kit only. ©Carolina Biological Supply Company disclaims all responsibility for any other uses of these materials. Included in the kit are:

petri dishes (3 compartments)	bottle saline agar
dropping pipettes	vial green dye
vial red dye	vial barium chloride
vial potassium sulfate	vial bovine albumin
vial goat antibovine albumin	vial goat antihorse albumin
vial goat antiswine albumin	vial saline
plastic beaker	glass rod
understanding immunology booklets	

Also needed but not included in the kit is a small sample of raw hamburger.

METHODS

Your team should obtain a dropping pipette and a petri dish containing agar. The bottom of each section in the dish has been imprinted I, II, and III. Become familiar with the following procedures before beginning an exercise. Templates are shown in each exercise.

Making the Wells in Agar

1. Set the petri dish (right side up) over the appropriate template so the template and section match. Templates are shown in each exercise.
2. Remove the dish cover and hold the dropping pipette vertically over one of the circles on the template.
3. Squeeze the pipette bulb and gently touch the pipette tip to the surface of the agar. While releasing the bulb, push the pipette tip down through the agar to the bottom of the dish. Lift the pipette vertically; this should leave a straight walled well in the agar.

 Caution

As the bulb is released, a vacuum pull is exerted on the agar plug. If this vacuum is not maintained while pushing the pipette into the agar, hairline fractures develop in the well that will interfere with the results.

Filling the Wells

1. Each vial has a dropper top. Do not exchange the droppers or use the dropper of one vial for solution in another vial. Draw a small amount of solution into the dropper, avoiding air bubbles.
2. Wipe the dropper tip on the inside edge of the vial to remove any excess solution from the tip.
3. Insert the dropper tip to the bottom of the appropriate well and slowly eject solution until the well is filled to, but not above, the surface of the agar. Either overfilling or underfilling a well may cause poor results. Immediately return the dropper to its vial.

Exercise 1: Diffusion and Precipitation in Agar

1. Set petri dish Section I over Template I.
2. Make the four wells indicated and then fill them as explained below.
3. Observe Section I at intervals for the next 30 to 45 minutes.

Well	Solution
1	Green dye
2	Red dye
3	Barium Chloride
4	Potassium Sulfate

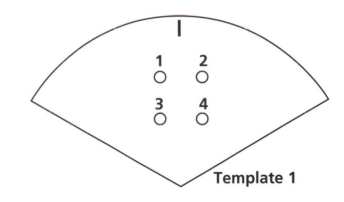

Template 1

DATA

At the end of this time, diagram your results on template I. (See "The Molecular Basis of Antigenic Specificity" and "Laboratory Applications of Immunology" in *Understanding Immunology* for background information.) The aforementioned Readings are included with the lab material. While waiting for results, proceed with Exercise 2.

RESULTS

What evidence of diffusion between Wells 1 and 2 do you observe?

Explain what has happened as barium chloride and potassium sulfate have diffused through the agar between Wells 3 and 4.

Exercise 2: Antibody-Antigen Reaction in Agar

1. Place petri dish Section II over Template II.
2. Make the four wells indicated and then fill them as explained below.
3. Replace the cover and set the dish at room temperature.
4. Observe your results 16 to 48 hours later. The results are best viewed by holding the dish, without its cover, vertically between the face and a light source, then moving the dish to the side until all glare vanishes.

Well	Solution
1	Bovine Albumin
2	Goat Antihorse Albumin
3	Goat Antibovine Albumin
4	Goat Antiswine Albumin

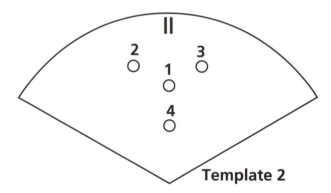

Template 2

DATA

Diagram the results on Template II. (See "Immunity" and "Antibody Production" in *Understanding Immunology* for background information.)

RESULTS

Which of the sera function as the antigen?

Which antibody has reacted with it?

If swine albumin had been placed in Well 1, where would the precipitin line be located?

Exercise 3: Testing for Food Purity

1. Set petri dish Section III over Template III and make the indicated four wells. Your instructor will prepare an extract of raw hamburger meat for testing.
2. Fill the wells as explained below.
3. Replace the cover and set the dish at room temperature.

Well	Solution
1	Hamburger extract
2	Goat Antihorse Albumin
3	Goat Antibovine Albumin
4	Goat Antiswine Albumin

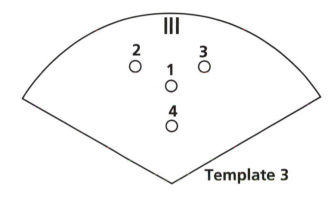

Template 3

DATA

Observe and record your results 16 to 48 hours later.

RESULTS

According to your test results, did the hamburger meat contain

beef?

pork?

horse meat?

Why is raw meat rather than cooked meat used in this test?

Lab 8

Food Nutrient Analysis

INTRODUCTION

Nutrients may be classified as carbohydrates (sugars and starches), lipids (fats and oils), proteins, and vitamins. In this exercise, sugar, starch, oil, protein, and vitamin C solutions will be tested. You will learn to test for nutrients and learn about common food items that contain them.

Carbohydrates are a group of organic compounds that contain atoms of carbon, hydrogen, and oxygen in the general chemical formula CH_2O_n. The biological process of respiration breaks down carbohydrates to yield energy for plant and animal activities. Carbohydrates also provide a source of carbon for use in the synthesis of many structural materials. Based on the complexity of molecular structure, carbohydrates may be divided into three broad categories: monosaccharides, disaccharides, and polysaccharides.

Monosaccharides are the least complex of the carbohydrate molecules and are often termed "simple sugars." Some examples of monosaccharides include glucose, fructose, ribose, and deoxyribose. A particular molecular structure of monosaccharides (the possession of free carbonyl group, $C = O$) enables them to reduce certain metallic ions (e.g., copper). Benedict solution, an alkaline solution of cupric sulfate ($CuSO_4$) can be used to detect monosaccharides. In the presence of a monosaccharide and heat, Benedict solution changes from light green to yellow to orange. These color changes result from the reduction of divalent Cu^{++} ions to monovalent Cu^+ ions, which then form a red compound, cuprous oxide (CuO_2).

Disaccharides are composed of two monosaccharide molecules chemically bonded together. Some examples of disaccharides include sucrose (table sugar), maltose, and lactose. The Benedict test detects only those disaccharides, such as maltose, which have a free carbonyl group. Sucrose and lactose give a negative Benedict test because the free carbonyl group is not present. However, if the bond between the monosaccharide molecules forming the disaccharide is broken, positive Benedict test results are obtained.

Polysaccharides, the most complex of the carbohydrates, are compounds of many monosaccharide molecules chemically bonded together. In nature, the majority of carbohydrates are polysaccharides and includes such compounds as cellulose, chitin, starches, and glycogen. The starches can be identified by a simple laboratory test using a solution of iodine-potassium-iodide. In the presence of starches, the iodine-potassium-iodide solution changes from bright yellow to dark blue. This color change results from the binding of iodine atoms (from the iodine-potassium-iodide) to the ends of the large starch molecules.

The lipids are a heterogeneous group of organic compounds classified together because they are insoluble in water, and soluble in organic solvents such as ether, benzene, and acetone. Lipids provide organisms with a concentrate source of energy, yielding twice as many calories per gram as carbohydrates or proteins.

The main source of lipids in the human diet is the triglycerides. They are composed of hydrogen, oxygen, and carbon atoms, and when chemically broken down, yield fatty acids and glycerol (an alcohol). At room temperature, triglycerides may be in solid or liquid form, depending on the nature of their fatty acids, and are termed fats or oils, respectively. If the fatty acids are mainly unsaturated (containing less than the maximum amount of hydrogen), the triglyceride will be oil. In general, fats are characteristic of animals and oils are characteristic of plants.

Lipids can be identified by a simple laboratory test using Sudan IV fat stain. Sudan IV is a dye, which is insoluble in water but is soluble in lipids. In a water and lipid (vegetable oil) mixture, the stain will be concentrated in the oil layer and will give it a scarlet color. The staining of lipids with Sudan IV is a physical process and, therefore, does not chemically alter the lipid molecules.

Proteins are complex organic compounds formed from long chains of amino acids chemically linked together. Amino acids contain atoms of hydrogen, oxygen, carbon, and nitrogen in the general chemical formula $NH_2–RCH–COOH$, where –R represents a carbon chain or ring that may carry –OH, –H, –S–, –P–, and the like. The properties of each amino acid depend on the structure of the R group.

All organisms require nitrogen for construction of proteins, nucleic acids, and certain other organic compounds. Plants and certain bacteria convert organic nitrogen into the amino group NH_2 that is used in the biosynthesis of amino acids. Twenty-three different amino acids are known to occur in living organisms. Any or all of these amino acids may occur in any number, sequence, and configuration to form an infinite number of proteins. Thus, the proteins and amino acids contained in foods are a source of prefabricated amino groups. Although some amino acids can be synthesized in the human body, 8 to 10 "essential amino acids" must be supplied in prefabricated form in the diet. The absence of any of these "essential amino acids" halts synthesis of certain proteins, which results in a diseased condition, and in prolonged cases, the eventual death of the organism.

The protein test used in this kit is called the biuret test because of the compound biuret that is formed in the presence of proteins. The test is positive for all compounds that contain two or more peptide bonds (bonds between amino acids). Thus, all proteins and polypeptides give a positive test, but individual amino acids do not. When solutions of sodium hydroxide and copper sulfate are mixed in the presence of a protein, a pink or purple precipitate forms, which on settling, leaves the liquid above it pink or purple. If the two solutions are mixed in the absence of a protein, a blue precipitate forms, which on settling, leaves the liquid above it clear.

In addition to the basic food groups (carbohydrates, lipids, and proteins), an organism's diet must also supply other substances for normal growth and health. These substances, required in small amounts and the lack of which result in deficiency diseases, are called vitamins. The vitamins do not share a common structure. They are usually divided into two groups: those that are water-soluble and those that are fat-soluble. Before the chemical nature of the individual vitamins was known, they were referred to by letters (A, B, C, and so on). This practice is still commonly followed today, although the chemical nature of the water soluble vitamins has been established and they have been given chemical names (thiamine for B1, ascorbic acid for C). However the fat soluble vitamins are more complex. For example, vitamin A is not one substance but several, all of which have the same nutritional function. Thus, it is not possible to give a single chemical name to vitamin A and other fat soluble vitamins.

Many vitamins are known to be incorporated into coenzymes and are necessary for the biosynthesis of other compounds. This is not known to be the case with vitamin C. The exact function of vitamin C in the body is unknown; however, it is probably active in tissue oxidations. Also, it is known to be present in the adrenal cortex and may participate in the biosynthesis of hormones in that gland. The vitamin C deficiency disease, scurvy, is characterized by bleeding gums and painful joints.

Vitamin C is widely distributed in nature, and is synthesized by many plants and all known mammals except the primates and the guinea pig.

The test for vitamin C in this kit depends on the fact that the vitamin is a strong reducing agent. Thus, as indophenol is reduced by the vitamin, the blue solution is bleached until it becomes colorless.

OBJECTIVE

Upon completion of this lab, you should be able to understand the tests available to detect the presence of sugars, starch, lipids, proteins, and vitamin C. In addition, you will have a better understanding of common food types that contain these nutrients.

SAFETY GUIDELINES

Comprehensive material safety data sheets are enclosed in the ©Carolina Biological Supply Company laboratory kits. Also, please review the safety guidelines found at the front of this manual. As always, you should take special precautions while using the boiling water baths.

 Warning
Some of the chemicals used in these exercises can stain skin and/or clothing.

MATERIALS

The materials that accompany the Food Nutrient Analysis BioKit® from ©Carolina Biological Supply Company #KZ20-2500 are supplied for use with the exercises in this kit only. ©Carolina Biological Supply Company disclaims all responsibility for any other uses of these materials.

glass vials	empty bottle
glass marking pencil	protein
Benedict solution	oil (vegetable)
starch	box of toothpicks
sugar (glucose)	Sudan IV Fat Stain (powder)
iodine-potassium-iodide solution	indophenol (0.1%)
sodium hydroxide solution (10%)	ascorbic acid
distilled water	four plastic measuring cups
copper sulfate solution (5%)	glass dropper

Other items needed are a hot water bath large enough to accommodate 10 vials. The hot water bath can be set up using a hot plate and a beaker, or a Bunsen burner, ring stand, ring clamp, asbestos wire, and a beaker.

METHOD

Sugar Test

1. With a glass marking pencil, label the vials 1 and 2, respectively. Also label them with your initials.
2. Measure 5 ml of water in the plastic cup provided and pour the water into vial 1.
3. To vial 2, add 5 ml of water and 10 drops of sugar solution.
4. Add 20 drops of Benedict solution to both vials.
5. Record your results under Data.

DATA

What color is the solution in vial 1?

What color is the solution in vial 2?

6. Carefully place both vials upright in the hot water bath and let them heat for 10 minutes. Watch the vials during this heating period and note any color changes in the solutions.
7. Remove the vials from the water bath.
8. Record your results under Data.

DATA

After heating, what color is the solution in vial 1?

After heating, what color is the solution in vial 2?

RESULTS

In this exercise, which solution is the control?

 Note

This test does not detect the presence of cane sugar (sucrose) and most other complex sugars.

Starch Test

1. With a glass marking pencil, label the vials 1 and 2, respectively. Also label them with your initials.
2. Measure 5 ml of water in the plastic cup provided and pour the water into vial 1.
3. To vial 2, add 5 ml of water and 10 drops of starch solution.
4. Add four drops of iodine-potassium-iodide solution to both vials. Watch the vials closely for 30 seconds.
5. Record your results under Data.

DATA

What color is the solution in vial 1?

What color is the solution in vial 2?

RESULTS

In this exercise, which solution is the control?

Lipid Test

1. With a glass marking pencil, label the vials 1 and 2, respectively. Also label them with your initials.
2. Measure 5 ml of water in the plastic cup provided and pour the water into vial 1.
3. To vial 2, add 5 ml of water and 10 drops of vegetable oil.
4. Use the end of a toothpick to transfer a small amount of Sudan IV fat stain to both vials.
5. Cover the open end of vial 1 with your thumb and shake the vial for about 5 seconds.
6. Shake vial 2 in the same manner.
7. Set the vials aside and do not move them for three minutes.
8. Record your results under Data.

DATA

What color is the solution in vial 1 at the end of three minutes?

A mixture of oil and water usually separates into two layers with the oil forming the top layer. What color is the top layer in vial 2 at the end of three minutes?

RESULTS

Does the Sudan IV appear to concentrate in the water or in the oil?

In this exercise, which solution is the control?

Protein Test

1. With a glass marking pencil, label the vials 1 and 2, respectively. Also label them with your initials.
2. Measure 5 ml of water in the plastic cup provided and pour the water into vial 1.
3. To vial 2, add 5 ml of water and 10 drops of protein solution.
4. Add 20 drops of sodium hydroxide to both vials.
5. Then add 20 drops of copper sulfate solution to both vials and gently shake the vials.
6. Record your results under Data.

DATA

What color is the solution in vial 1?

What color is the solution in vial 2?

RESULTS

In this exercise, which solution is the control?

Vitamin C Test

1. With a glass marking pencil, label the vials 1 and 2, respectively. Also label them with your initials.
2. Add 20 drops of indophenol solution to vial 1.
3. Then add 1% vitamin C solution one drop at a time, shaking the vial after the addition of each drop, until the indophenol solution becomes clear.
4. Record your results under Data.

DATA

How many drops of 1% vitamin C solution was needed to clear the indophenol solution?

5. Add 20 drops of indophenol solution to vial 2.
6. Then add water in drops to equal the amount of vitamin C solution that was added to vial 1.
7. Record your results under Data.

DATA

Does the addition of water cause the indophenol solution to clear?

RESULTS

In this exercise, which of the solutions is the control?

Reactions of some common foods:
Positive sugar test: honey, milk, fruit juices, and fruits including raisins.
Positive starch test: flour, bread, crackers, potatoes, and many processed foods that have corn starch added as a filler or binder.
Positive lipid test: cooking oils, bacon, pork, and other meats.
Positive protein test: egg white, milk, fish, meats, and dried beans.
Positive vitamin C test: citrus fruit juices (canned, frozen, or fresh), tomato juice, and the juices of several other vegetables such as cabbage.
A negative test does not indicate that a given food substance is absent from a food sample. However, a negative test result does indicate that a given food substance is not present in large quantities in the sample being tested.

Lab 9

Gram Stain and
Bacterial Morphology

LAB LINK TO VETERINARY PRACTICE

You are at a farm doing reproductive exams and are prepared to leave when Mrs. Farmer asks you to look at one more cow. When she milked the cow this morning, she noticed that one quarter seemed much harder and the cow just wasn't eating well.

As you perform the physical examination, you detect a fever of 105°, 3° above normal. While examining the udder, you detect that the left rear quarter is swollen, hot to the touch, and red. The milk from this quarter is slightly off-color with abnormal lumps and flakes in it. You tell Mrs. Farmer that it is an obvious case of mastitis, but you are not sure what organism is causing the problem. You take a sample from the quarter to use for culturing. While there, you direct Mrs. Farmer to administer a medication to help relieve the fever and choose an antibiotic based on your experience.

After returning to the office, you set up the culture of the milk sample to identify what specific organism is infecting the quarter. Because the culture results will not be available for 12 to 24 hours, you do a Gram stain of the sample in an attempt to identify what class of organism is present. Because treatment is very different between Gram positive and negative organisms, this information can be extremely valuable.

INTRODUCTION

If the human eye could resolve images as well as the light microscope, we would see bacteria and fungi virtually everywhere. They grow in air, water, foods, and soil, as well as in plant and animal tissue. Any environment that can support life has its bacterial or fungal population.

Bacteria and fungi affect humans in various ways. Some cause animal diseases such as abscesses, tuberculosis, mastitis, and ringworm. Many bacteria and fungi also affect humans and plants. Most microorganisms do animals and humans little or no harm, and many are vital to our well-being and continued existence on earth. Bacteria and fungi are involved in the recycling of matter, purification of sewage, and filtration of water in the soil. They are essential to the production of cheeses, sauerkraut, pickles, alcoholic beverages, and breads. Biotechnology firms use microorganisms to produce antibiotics, amino acids, interferons, enzymes, and human growth hormones.

Bacteria and fungi are convenient organisms for research in genetics, physiology, cytology, and biochemistry because they grow rapidly, are easy to manipulate, and require only minimal laboratory space compared to mice or guinea pigs. As prokaryotes, bacteria have the advantage of being relatively simple organisms. On the other hand, fungi, which are eukaryotes and thus much more complex genetically, grow so quickly that a number of generations can be obtained in only a short period of time.

This laboratory exercise should train you in the extremely valuable technique of performing the Gram stain. This procedure, which is used in classifying bacteria, is also extremely valuable in guiding treatment.

OBJECTIVE

Upon completion of this lab, you should be able to perform correct Gram staining technique for bacteria and explain the morphology of bacteria.

SAFETY GUIDELINES

Comprehensive material safety data sheets are enclosed in the ©Carolina Biological Supply Company laboratory kits. Also, please review the safety guidelines found at the front of this manual. Use extreme caution while utilizing the flame source and protect hands from staining with disposable gloves.

MATERIALS

The materials that accompany the Gram Stain and Bacterial Morphology Kit from ©Carolina Biological Supply Company #KZ-15-4724 are supplied for use with the exercises in this kit only. ©Carolina Biological Supply Company disclaims all responsibility for any other uses of these materials. Included in the kit are:

Bacillus megaterium tube culture (Gram + rods)
Escherichia coli tube culture (Gram – rods)
Micrococcus luteus tube culture (Gram + cocci)
Rhodospirillum rubrum tube culture (Gram – spirals)

disposable inoculating loops
ethanol 95%
microscope slides
Gram iodine
safranin counterstain
autoclave disposable bag
Hucker ammonium oxalate crystal violet

⚠ Note

It is ideal to use the cultures within 24 hours. As the cultures age, the staining characteristics will change, making Gram + bacteria appear Gram –.

Materials needed but not included in the kit are:

Bunsen burner or similar flame source
distilled water
microscope with oil immersion capability
clothespins

immersion oil
lens paper for cleaning the microscope objectives
disposable gloves to protect the hands while staining

METHODS

Aseptic Technique

In most microbiological procedures, it is necessary to protect instruments, containers, media, and stock cultures from contamination by microorganisms constantly present in the environment. Aseptic technique involves the sterilization of tools, glassware, and media before use, as well as measures to prevent subsequent contamination by contact with nonsterile objects.

Before working with bacterial or fungal cultures, always wash your hands with soap and water. Next, prepare a work area. Select an area that is as free from drafts as possible. Turn off the air-conditioner and fans, and close all windows and doors. Wipe the work area with 70% ethanol or similar disinfectant solution. Arrange your materials conveniently on the clean work surface. Do not smoke, eat, or drink while working with cultures.

Following all work, the area should once again be thoroughly cleaned with an appropriate disinfectant. Contaminated wastes should be autoclaved or incinerated. If an accidental spill or breakage occurs, do not panic, but consult your instructor for appropriate guidance.

Gram Stain Technique

Morphology. Bacteria vary greatly in size, but their cell shapes are of three basic types: coccus (sphere-shaped), bacillus (rod-shaped), and spirillum (spiral or comma-shaped). Some bacteria exist singly while others are attached in chains or packets.

Bacterial cells can be colored with a stain to provide contrast with the background or to make cellular organelles visible. Differential stains such as the Gram stain are more complex and are used to divide bacteria into groups. Bacteria stain differentially because they differ in cell wall composition. The Gram stain separates almost all bacteria into two large groups: the Gram positive bacteria, which stain blue, and the Gram negative bacteria, which stain pink. This classification is basic to bacteriological identification.

Staining

1. Place a drop of distilled water on a clean slide.
2. Remove a small quantity of bacteria from the culture media using the disposable inoculating loop.
3. Flame the mouth of the tube and replace the cap. This technique is performed by quickly passing the mouth of the glass tube through the flame of the Bunsen burner several times. Do not perform this step if a plastic tube is used.
4. Mix the bacteria with the water on the slide and spread into a thin film. It is essential to keep this side of the slide on the up side throughout the entire process.
5. Allow the smear on the slide to air dry.
6. Using a clothespin or similar holding device, pass the slide, smear side up, through a flame three times to fix the bacterial cells. Fixing kills the bacteria and causes them to stick to the slide.
7. Allow the slide to cool.

⚠ Note

During the staining procedure, continue to use the clothespin to hold the slide to minimize the risk of getting stain on your hands. Disposable gloves are very helpful in this procedure. Protective clothing or "old" clothing could be worn so that a stain spill does not destroy valuable clothing.

8. Flood the smear side of the slide with Hucker ammonium oxalate crystal violet for 60 seconds. If the slide is held level and steady, the stain can be applied to the slide and held there for 60 seconds. More stain needs to be applied only if the stain spills from the slide.
9. Rinse with tap water.
10. Flood with Gram's iodine solution for 60 seconds.
11. Rinse with tap water.
12. Decolorize with 95% ethanol. Allow the ethanol to drip across the slide until the runoff is almost clear. Do not over decolorize as this will eliminate the color from Gram positive organisms.
13. Rinse with tap water.

14. Flood with safranin for 60 seconds. Safranin is considered the counterstain.
15. Rinse with tap water.
16. Blot carefully and gently. Do *not* wipe. Allow to dry completely.
17. Examine the slide under oil immersion. As always, begin with a low-power objective and get the slide in focus. Then move step by step to each higher power objective. Use only the fine focus knob when using the oil immersion lens. This lens can easily be damaged if the stage is brought down too hard against the lens.
18. Repeat the procedure for all of the bacteria types.
19. When finished with the microscope, be sure to clean the lens thoroughly with appropriate lens paper. It is important to use cleaning materials that are designed for the delicate optics of the microscope.

DATA

Record your findings on the size, shape, grouping, and staining characteristics of each bacteria.

Bacillus megaterium

Escherichia coli

Micrococcuss luteus

Rhodospirillum rubrum

RESULTS

Gram positive organisms—blue
Gram negative organisms—red

Lab 10

Bacterial Culturing and Inhibition

LAB LINK TO VETERINARY PRACTICE

Tex T. Rainer from the local racetrack calls your office and asks to speak to you. He tells you that his horse runs too slowly. He claims the horse even seems to breathe heavily while it stands in the stall. You promise to be at the track this afternoon to take a look.

When you arrive, it is obvious that the horse is breathing heavily, his ears are drooped, and he even has a small amount of thick nasal discharge. On physical examination, you discover that his temperature is elevated to 102.4°F (you expect the normal to be closer to 100.5). Using your stethoscope, you detect abnormal lung sounds representing an increased amount of fluid within the airways.

You explain to Tex that the horse has pneumonia and that a course of antibiotics is necessary. To ensure that you have chosen the correct antibiotic, you elect to perform a transtracheal wash. In this procedure, you insert a catheter that passes through a needle deep into the trachea. You flush in a small amount of sterile saline and quickly draw back as much of the fluid as possible. You explain to Tex that you are going to take this sample back to the office for further testing.

At the office, this sample is used to culture the pathogenic bacteria. Once the organism is identified, an antibiotic sensitivity test is performed to establish which antibiotics are appropriate for treatment. You select the antibiotic that is most likely to be successful and begin treatment while you wait for culture results.

INTRODUCTION

Bacterial diseases are quite common in veterinary and human medicine. Proper bacterial culturing and antibiotic sensitivity testing are essential techniques used to properly diagnose and treat infections.

OBJECTIVE

Upon completion of this lab, you should be familiar with the proper techniques in culturing bacteria and assessing sensitivity to antibiotics and disinfectants.

SAFETY GUIDELINES

Comprehensive material safety data sheets are enclosed in the ©Carolina Biological Supply Company Microbiology Apparatus laboratory kits. Also, please review the safety guidelines found at the front of this manual. Handle all cultures of microorganisms carefully, especially those known to be pathogenic. In other words, consistently practice good aseptic technique as outlined below. Consult the instructions for bottle media found in the appendix for guidance on proper disposal of supplies. Guidelines vary between states, municipalities, and schools; therefore, it is important to check for local requirements.

MATERIALS

The materials that accompany the laboratory kits from ©Carolina Biological Supply Company #KZ76-6200 are supplied for use with the exercises in this kit only. ©Carolina Biological Supply Company disclaims all responsibility for any other uses of these materials. Included supplies are:

- bacterial type slide
- inoculating needle
- sterile petri dishes
- ethanol 95% denatured
- sterile pipettes
- microscope slides
- alcohol lamp
- nutrient agar bottle medium
- dropper bottle stain
- disposal bags

METHODS

Aseptic Technique

To avoid contamination from airborne organisms during medium preparation and inoculation, stop all unnecessary air movements (doors, windows, fans, and the like). Wipe all work surfaces with a disinfectant (e.g., alcohol, 5% household bleach, or Lysol®). Wash hands before and after each work session. Do not eat or put anything in your mouth while working in the laboratory. Consult the instructions for bottle media found in the appendix for appropriate disposal procedures.

Preparing Plates

The agar plates can be prepared before class or at the beginning of the class session. If the plates are prepared at the beginning of the period, they will have cooled sufficiently for inoculation before the class period is over. To prepare the media, consult the instructions for bottle media found in the appendix. Pour all plates.

Collecting Microorganisms from the Environment

Air: Uncover petri dishes for 20 minutes in various areas such as the classroom, outdoors, and so on.

Water: With a sterile pipette, spot five to 10 drops of water on the plate. Try several sources of water to sample. Disposable cups can be used to collect water samples since they are nearly sterile when first opened.

Person: Cough on open plate; streak fingertip across open plate; put hair on open plate; streak cotton swab between teeth and streak on plate.

Other: Expose open plate to direct contact with coins, lint, soil, dust, animal fur, feathers, and so on.

Observing Colonies and Isolating Pure Cultures

Various microorganisms grow in one or two days when incubated at room temperature. Colonies may show a wide variety of forms. Consult references such as bacteriology manuals or the Internet to try to identify some of the bacteria you have collected. To isolate a pure culture, follow the streaking technique found in the instructions for bottled media found in the appendix. If a distinct colony is found, collect a sample on a sterile wire loop and reisolate it on another petri dish, following the same streaking pattern. This time, if the isolation succeeds, only one colony of only one type will be present. Assume this to be a pure culture.

Slide Preparation

For a fast and simple stain, touch a small portion of a colony with an applicator stick. Smear this on a glass microscope slide. Add two drops of water and spread again. Allow this mixture to air dry. Add several drops of 0.1% crystal violet and wait for 30 seconds. Wash the excess off gently. Place the slide on end to allow it to air dry. With the oil immersion lens, examine the prepared slide.

A more involved process would be to perform the Gram stain technique. This technique discussed in Lab 9, Gram Stain and Bacterial Morphology, will help to further classify the bacteria.

DATA

Draw what you found on the slides.

RESULTS

Are the bacteria Gram positive or Gram negative organisms?

What shape bacterium is it? (cocci, bacillus, spiral)

Are the bacteria individual or clustered into groups?

BACTERIAL INHIBITION

INTRODUCTION

The Bacterial Inhibition Kit from ©Carolina Biological Supply Company #KZ76-6250 introduces you to the study of the effects of antibiotics and disinfectants on a microorganism.

ADDITIONAL SAFETY GUIDELINES

You may handle the control disks and the antibiotic disks with your fingers. The accidental introduction of stray microorganisms into the plates at this time with the heavy lawn of *Bacillus cereus* is not likely to adversely affect the outcome of the exercise.

MATERIALS

The materials that accompany the Bacterial Inhibition Kit is available from ©Carolina Biological Supply Company #KZ76-6250 are supplied for use with the exercises in this kit only. ©Carolina Biological Supply Company disclaims all responsibility for any other uses of these materials. Included in the kit are:

petri dishes	nutrient agar
control disks	antibiotic disks
disinfectant disks	*Bacillus cereus* suspension

Other items needed are:

incubator or warming area set at 40°C	an appropriate disinfectant (alcohol
cleaning supplies	5% household bleach or other disinfectant)

 Note
Cleanliness is essential when working with microorganisms. It is important to wash hands and wipe all working surfaces with a disinfectant (alcohol, 5% bleach, or soap and water) before and after each session. Put nothing in your mouth.

Seeding Plates

Your team will receive four prepared plates of nutrient medium. These plates should be labeled 1, 2, 3, and 4. The instructor will pour about 1 ml of a culture of

Bacillus cereus onto the medium in plate 1. Immediately swirl the plate gently to ensure that the entire surface of the medium is covered by the culture, then pour the excess into plate 2. Repeat this procedure without delay on plates 3 and 4. Any excess on plate 4 should be poured into the laboratory sink when indicated by your instructor. During this procedure, the covers of the dishes should be lifted no more than necessary and should be replaced on the plates immediately.

Control Disks

One untreated control disk should be placed off-center on the surface of the medium of each of the four plates before you handle either the antibiotics or disinfectants.

Antibiotic Disks

A different antibiotic will be tested on each of plates 1, 2, and 3. Two disks of one antibiotic should be placed on the surface of the medium of each plate. The disks must be placed on the medium before the bacteria have begun to grow.

Disinfectants

Plate 4 will be used to test two disinfectants, one disk of each type. Use a separate color for each disinfectant. Using forceps, pick up a disk and saturate it with disinfectant by capillary action. Place the saturated disk on the seeded plate. Record the disinfectant and its color on your Data sheets below.

Incubation

Incubate plates until sufficient growth appears to permit detection of zones of inhibition. An incubation temperature of 35°C to 37°C yields quickest results, but room temperature works fine. A zone of inhibition around the disk indicates that the organism is sensitive to the specific agent tested. You should be able to observe this zone in one or two days. Measure the diameter of the zone of inhibition and record the results in the following chart.

The plates should not be opened for examination. Once the disks have been placed on the plates, the cover should be kept on the dish and not removed. The bacterium supplied is not pathogenic, but any culture should be handled as a potential pathogen; in other words, use good aseptic technique.

DATA

Plate #	Treatment	Results (Diameter of Zone)	
		Treated	**Controls**
1. Antibiotic			
2. Antibiotic			
3. Antibiotic			
4a. Disinfectant			
4b. Disinfectant			

ADDITIONAL OBSERVATIONS

Write down any other findings you noted.

RESULTS

Following incubation, a clear area surrounding a disk indicates the microorganisms have been inhibited by the treatment. You should be careful not to make broad generalizations about the effectiveness of the materials studied in this exercise. You have tested the materials against a single species *Bacillus cereus*. An antibiotic or disinfectant that is effective against one microorganism may not be effective against another microorganism. The biological effect of an antibiotic or a disinfectant on any specific microorganism is influenced by its concentration and the degradation of the material occurring since it was first prepared. Thorough testing of an antibiotic or a disinfectant would include many different pathogenic bacteria at various concentrations of the antibiotic, as well as testing the antibiotic after storage under various known conditions to determine the rate of degradation.

In the space provided below, answer the following questions. When interpreting the results, consider whether they were as anticipated.

Was one antibiotic more effective than another?

Was one disinfectant better than another?

Addendum

Epidemiology Lab

LAB LINK TO VETERINARY PRACTICE

Assume you are an epidemiologist for the department of agriculture. On Monday afternoon, you receive a phone call from the state veterinarian that a commercial poultry flock in your district has tested positive for a very virulent strain of avian influenza. You phone the distraught producer and find that a class of veterinary science students visited the farm one week ago, also on a Monday. Following up on this information, you contact the classroom instructor only to discover that the class had visited a live bird market in New York City the Friday before the poultry farm visit.

Using deductive reasoning, you surmise that the students were in contact with some infected birds at the New York City venue and likely transferred the disease to the local poultry operation one week ago today. It is now your job to track other potential poultry flocks the students may have contacted since the New York City trip.

You also know that avian influenza does not necessarily need bird-to-bird contact for infection to take place. The virus may be harbored in or on infectious persons for up to seven days and can be transferred to others who work with birds.

INTRODUCTION

The field of epidemiology includes tracking and preventing disease outbreaks. As an epidemiologist, skill at tracking possible routes of disease transfer is critical to preventing diseases from crippling the United States poultry industry.

OBJECTIVE

Pretending it was your class that visited the live bird market and the poultry operation, students will track all the potential ways the individuals from the class may have transferred avian influenza to other poultry flocks or wild birds during the intervening week between the New York trip and the poultry farm visit.

SAFETY GUIDELINES

None

MATERIALS

For this exercise, copies of a local (county) map, a state map, and a national map to track the routes of disease transfer are needed.

METHODS

Independently, each student should mentally trace his or her physical movements for the seven days between the New York visit and the poultry farm visit. Students should note each potential contact with domestic poultry, wild birds, pet birds, or other people who may have direct avian contact. After students have completed this exercise, compile all potential routes of disease transfer and subsequent downstream disease transmission within the week on a state, local, or (if necessary) national map. The resulting map should resemble an airline routing map with potential transmission routes leading in various directions from the school.

DATA

	Transferred to	Transferred to	Transferred to
Contact #1			
Contact #2			
Contact #3			
Contact #4			
Contact #5			
Contact #6			
Contact #7			
Contact #8			

Add more rows as necessary.

RESULTS

How many counties or states did your class identify as potential recipients of avian influenza from your class visit to New York?

What measures could be taken to prevent disease spread, or to track potential routes of disease transfer?

Appendix

Instructions for Bottled Media

> ⚠ Caution
>
> Do not attempt to melt bottled media in the microwave. It is not possible to vent the bottle without contaminating the media. Heating the bottle in the microwave may cause pressure buildup, thereby creating an explosion hazard. For your safety, please follow the recommended procedures below.

PREPARING THE MEDIUM

1. Fill a small pot halfway with water (or use a large pot for several bottles).
2. Place the bottle(s) of medium unopened inside the pot. Loosen the cap slightly. Adjust the water level to make sure it is even with the level of the agar in the bottle(s).

> ⚠ Caution
>
> Removing the cap from the bottle before pouring the plate will contaminate the medium. Do not place a cold bottle into boiling water! It may cause the glass bottle to shatter.

3. Heat the pot until the medium has melted completely. It may take up to 30 minutes to completely melt the medium.
4. Using an oven mitt or a pot holder, remove the bottle(s) from the water and set on a heat-resistant surface.

> ⚠ Caution
>
> In addition to being hot, the bottle(s) may be slippery. Please use extreme caution.

5. Allow the bottle(s) to cool to approximately 45°C before pouring plates. At that point, the medium will still be liquid but the bottle(s) will not be too hot to handle.

> ⚠ Note
>
> Do not place the bottle(s) in an ice bath or refrigerator in an attempt to speed the cooling process.

PREPARING THE WORK AREA

1. Sanitize and disinfect the work area by cleaning with 70% alcohol or a bleach solution. Allow the area to air dry.

Note

A bleach solution may be prepared by adding 4 parts water to 1 part household bleach. Be careful with this solution as it may damage clothing, and is potentially damaging to skin and eyes.

2. Wash and dry your hands thoroughly.

POURING THE PLATES

1. Place the agar plates on the prepared work surface.
2. Remove the cap of the bottle you are going to use. Sterilize the mouth of the medium bottle by passing it through the flame of a Bunsen burner or by wiping with an alcohol prep.
3. Lift the cover of a plate just enough to pour in the medium.
4. Pour a small amount of agar medium into the plate. Use just enough to cover the bottom of the dish. Each bottle will cover approximately five 100 mm plates.

 Note

Pour only one plate at a time to minimize contamination. Hold the agar bottle tilted at a 45° angle until all the medium in the bottle has been used to pour plates. Never set the bottle down or hold it upright during the pouring process because this increases the potential for contamination.

5. Immediately close the plate cover. Remember: Do not set the bottle down or hold it upright.
6. Follow the same steps to pour the remaining plates.
7. Allow at least 30 minutes for the plates to set up.
8. Dispose of the empty bottles (see Disposal section).

INOCULATING THE STREAK PLATES

Prior to inoculating the streak plates, it is important that the agar surface be smooth and moist; however, excessive moisture is undesirable since it can cause the bacterial colonies cultured on the plate to merge. Be sure the agar is allowed to set for at least 30 minutes before establishing cultures on the plates.

 Note

For a standard inoculation, streak in the simple zigzag pattern shown in Figure A–1 using either a sterile swab or wire loop. If you are trying to isolate types of bacteria on a plate, use an isolation streaking pattern such as the one shown in Figure A–2. Isolation streaking requires a wire loop, which must be flamed between each of the streaks on a plate. In isolating bacteria, each successive streak creates a more "diluted" population on its section of the plate.

1. Place the plates on the clean work surface.
2. Use either a fresh, sterile, cotton swab just removed from its package or a thoroughly flamed wire loop.

Figure A–1. Simple streak (use sterile loop or swab)

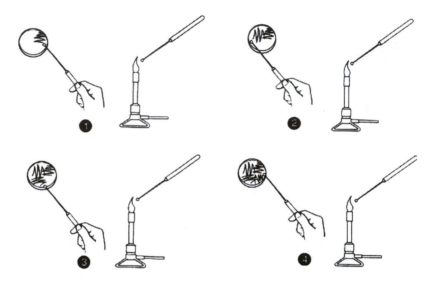

Figure A–2. Isolation streaking (use metal loop only and flame it between streaks)

⚠ Note

You must use a new, sterile swab or a flame-sterilized loop for each plate. Be sure to allow the loop to cool following the flaming. If the loop is immediately introduced into the bacteria, the heat may kill the bacteria and/or damage the culture media.

3. Rub the swab or loop over, or place it in, the material to be cultured.
4. Lift the lid of an agar plate.
5. Streak the plate. Use either a simple zigzag streak (Figure A–1) or an isolation streaking method (Figure A–2). Quickly replace the lid after making each streak. Do not press too hard while streaking the plate. Too much pressure will damage the surface of the medium.
   Remember: For isolation streaking, a wire loop must be used.
6. Incubate the plates.

Note

For room temperature incubation, place the plates agar-side up. If using an incubator, invert the plates and incubate them at 30°C to 37°C. The specific time and temperature required for incubation depends on the medium selected.

7. Sterilize wire loops or dispose of used swabs (see Disposal section).

STORAGE

Unopened medium bottles may be stored under refrigeration for up to one year or at room temperature for up to six months.

DISPOSAL

For used plates, medium, bottles, and swabs.

Method 1

1. Double-bag all used or contaminated materials with autoclavable disposable bags.

 Note

Contaminated refers to any package whose original seal was broken, even if the contents were not used.

2. Vent the disposal bag to allow steam to penetrate the inner contents.
3. Place in an autoclave at 121°C at 15 psi for 45 minutes.

 Note

Time may vary from state to state. Always check with the public health service in your state to determine the requirements for decontamination.

4. Indicate on the bag with special autoclave indicatory tape that the enclosed sample has been autoclaved.
5. Dispose of in an appropriate waste receptacle.

Method 2

1. Soak the used or contaminated material in a bleach solution (1 part household bleach to 3 parts water) for 24 hours.
2. Place the materials in the disposal bag and seal it.
3. Incinerate the bag.
4. For wire inoculating loops, flame the wire in a Bunsen burner to sterilize it.

Tips for Successful Cultivation

1. Avoid contaminants. Do not use nonsterile plates, swabs, or media.
2. Properly sanitize and disinfect the work area according to instructions.

3. Label and date each individual petri dish. This will make recordkeeping much easier.
4. Use the proper inoculation techniques for specific culture media in order to obtain the correct biochemical growth patterns or reactions.
5. Use the correct medium for cultivation.
6. Incubate under appropriate conditions.
7. Follow preparation and disposal procedures.

Delmar Cengage Learning Order Form

Lab No.		Item	Price
1	Beginner's Introduction to Cells Kit	KZ-31-9826	$24.85
2	Beginner's Animal Tissue Slide Set	KZ-31-1956	$65.00
3	ATP-Muscle Kit	KZ-20-3525	$41.00
4	Urine Exam Biokit	KZ-69-5830	$82.75
5	Digestion Biokit	KZ-20-2340	$124.00
6	Cellulose Biokit	KZ-20-2345	$71.95
7	Starch Breakdown Kit	KZ-20-2335	$84.00
8	Diabetes Basic Kit	KZ 70-0440	$69.90
9	Immunodetective Biokit	KZ-20-2100	$154.75
10	Food Nutrient Analysis Biokit	KZ-20-2500	$64.25
11	Gram Stain and Bacterial Morphology Kit	KZ-15-4724	$62.50
12	Bacterial Inhibition Kit	KZ-76-6250	$64.25
12	Microbiology Apparatus Kit	KZ-76-6200	$69.50

**Carolina Biological Supply Company   Phone: 800 334 5551   email: carolina@carolina.com
Fax: 800 222 7112   www.carolina.com**

Notes

Notes

Notes

Notes

Notes

Notes

Notes

Notes

Notes

Notes

Notes

Notes

Notes

Notes